Dinosaur Publicatior

Caterpillars to Moths

by Althea

illustrated by Maureen Galvani

Published by Dinosaur Publications Ltd Over Cambridge Great Britain

© Dinosaur Publications 1983
© text Althea Braithwaite 1983
© illustrations Maureen Galvani 1983

ISBN 0/85122/358-3 (paperback)
ISBN 0/85122/359-1 (hardback)
Made in Great Britain

Unlike butterflies, most moths fly at night. Moths are attracted by light but blinded by it too. When they fly in through an open window, they bump into walls and furniture. They also bump into people and this may be why some people don't like them!

There are many more kinds of moths than butterflies in this country. Some are very beautiful.

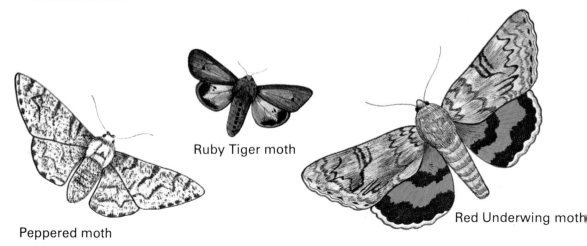

Ruby Tiger moth

Peppered moth

Red Underwing moth

Moths usually have fatter, furrier bodies than butterflies, but the surest way to tell them apart is by their antennae. Butterflies always have antennae with thickened, club-shaped ends.

Antenna of Butterfly

Moths have pointed or blunt antennae, many are feathery too.

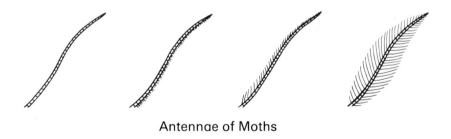

Antennae of Moths

The hind wings of moths have tiny bristles which catch on to a sort of loop on their front wings. This hooks their wings together when flying.

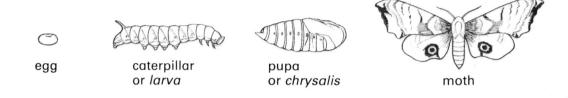

egg

caterpillar
or *larva*

pupa
or *chrysalis*

moth

Like butterflies, moths go through four different stages in their lives. The caterpillar eats and eats and keeps getting too big for its skin. Each time the skin splits a new skin has grown underneath. Some caterpillars hibernate for the winter, waking in the spring to eat some more.

When the caterpillar is fully grown the last skin splits, and there inside is the pupa. Moth caterpillars often spin a cocoon around themselves, and some bury themselves under the ground before they pupate.

It is difficult to recognize a moth caterpillar by its choice of foodplant, as many of them eat a wide range of different plants.

Square Spot Rustic moths

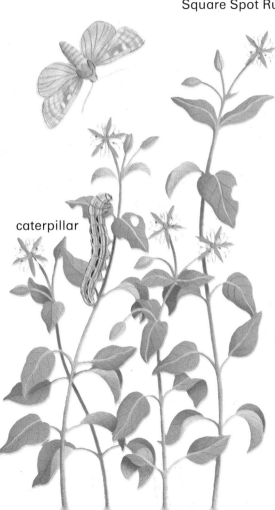

caterpillar

Square Spot Rustics are very common in most parts of Britain. They come to lighted windows at night and are often eaten by bats who snap them up as they fly. These are the dull-coloured moths that many people think of when they talk about moths.

The caterpillar feeds at night on dock, chickweed or grasses.

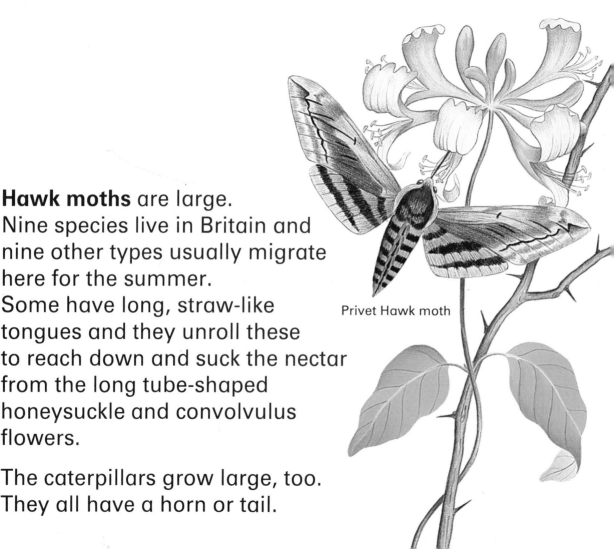

Hawk moths are large.
Nine species live in Britain and
nine other types usually migrate
here for the summer.
Some have long, straw-like
tongues and they unroll these
to reach down and suck the nectar
from the long tube-shaped
honeysuckle and convolvulus
flowers.

The caterpillars grow large, too.
They all have a horn or tail.

Privet Hawk moth

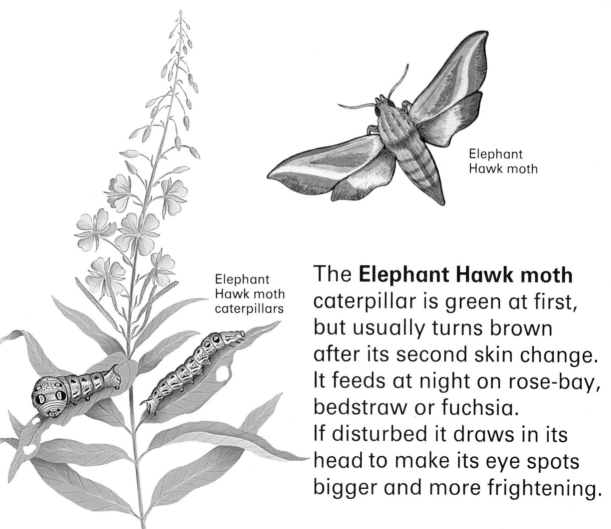

Elephant
Hawk moth

Elephant
Hawk moth
caterpillars

The **Elephant Hawk moth** caterpillar is green at first, but usually turns brown after its second skin change. It feeds at night on rose-bay, bedstraw or fuchsia.
If disturbed it draws in its head to make its eye spots bigger and more frightening.

The caterpillars of both the **Eyed Hawk moth** and the more common **Poplar Hawk moth** feed on sallow and willow, as well as other trees. The Eyed Hawk moth takes about six nights to lay her 400 eggs, one or two at a time.

When resting, both kinds of moth look like dead leaves, but if frightened the Eyed Hawk moth displays its hind wings with two large eye spots.

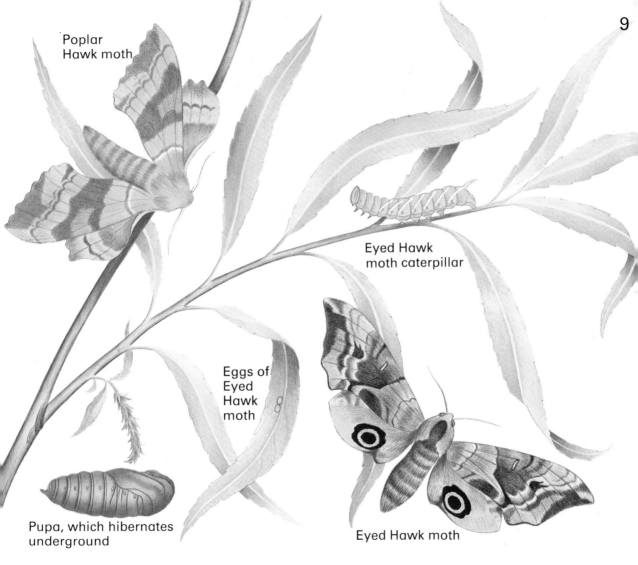

Poplar
Hawk moth

Eyed Hawk
moth caterpillar

Eggs of
Eyed
Hawk
moth

Pupa, which hibernates
underground

Eyed Hawk moth

The **Puss moth** is named after its large, fluffy, cat-like body. The caterpillar feeds on sallow, willow or poplar leaves. If disturbed it makes itself look frightening by rearing up and drawing in its head, showing a red face with two black eye spots. It also lashes out with two red filaments from its forked tail.

When fully grown, in August or September, the caterpillar gnaws a small hollow in the bark of the tree and, using the chewed-up wood, makes a shell-like cocoon. The moth cuts its way out of the cocoon the following May.

The **Sallow Kitten** and the **Poplar Kitten** moths are so named because they are smaller than the Puss moth.

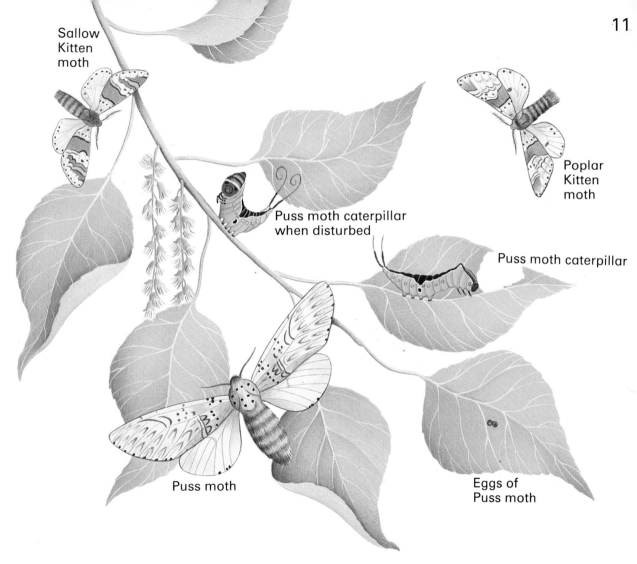

Sallow Kitten moth

Poplar Kitten moth

Puss moth caterpillar when disturbed

Puss moth caterpillar

Puss moth

Eggs of Puss moth

The tiny black caterpillars of the **Emperor moth** feed together on heather, bramble or sallow. As they grow bigger, they change colour and go off on their own to feed.

This moth belongs to the silk moth family and, before pupating, the caterpillar spins a tough, brown silk cocoon. The cocoon has a pointed end with a ring of spikes around it which is pushed open from the inside when the moth emerges the following spring.

The female is much larger than the male, and they each have large eye-spots on all four wings.

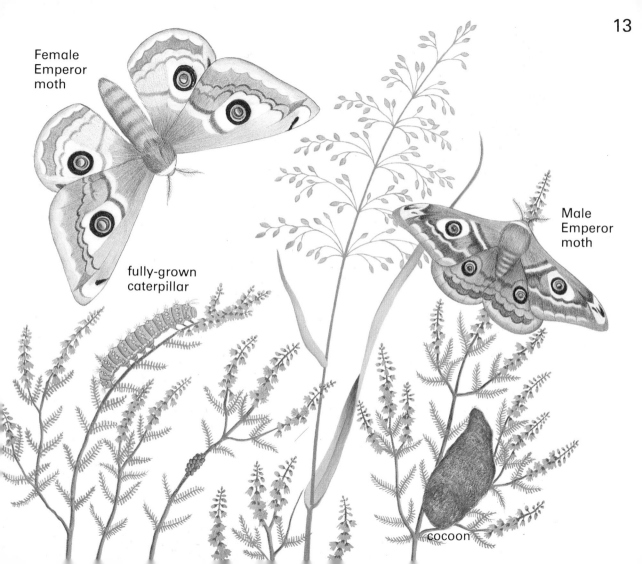

Female
Emperor
moth

Male
Emperor
moth

fully-grown
caterpillar

cocoon

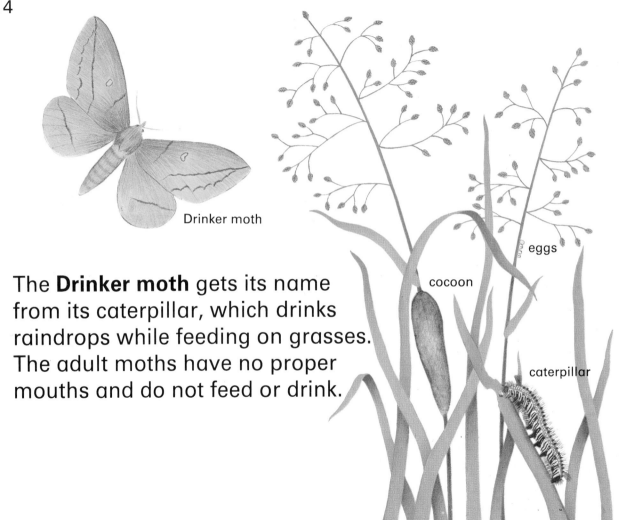

Drinker moth

eggs

cocoon

caterpillar

The **Drinker moth** gets its name from its caterpillar, which drinks raindrops while feeding on grasses. The adult moths have no proper mouths and do not feed or drink.

The **Oak Eggar moth** belongs to the same family as the Drinker. After mating, the female flies around dropping her eggs.
Luckily, the caterpillar can feed on many plants including bramble, blackthorn, broom and heather.
If disturbed, the caterpillar will fall off the foodplant and roll up into a ring.

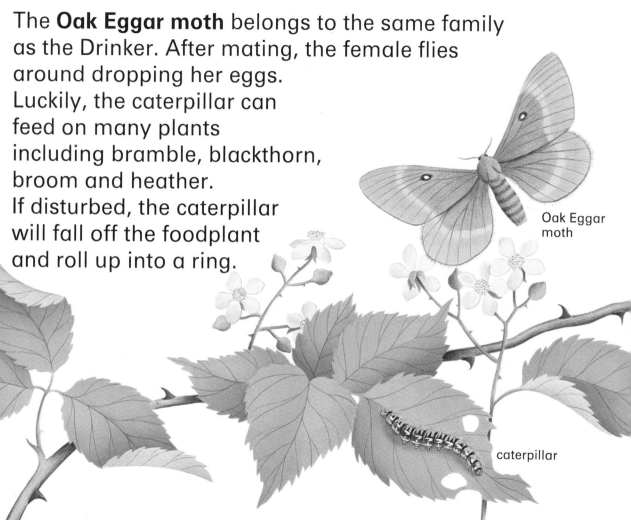

Oak Eggar moth

caterpillar

Cinnabar caterpillars hatch in dozens
from each shiny, yellow egg. They feed together
on ragwort, so if you find one, you will soon
notice lots more. They pupate just under
the ground.

Unlike most moths, the upper and undersides
of its wings are coloured the same.
They fly during the day as well as at night.
Because their bright colours warn birds that
they are poisonous, Cinnabars are not
frightened of being attacked.

caterpillars

Cinnabar moth

The hairy caterpillars of the **Garden Tiger moth** feed on many plants including nettles and dandelions. Most birds, except the Cuckoo, avoid eating hairy caterpillars as the hairs can irritate their throats.

This moth flies at night, but does not bother to hide in the daytime. It can frighten attackers by opening its red hindwings, and exposing the fringe of red hairs behind its head.

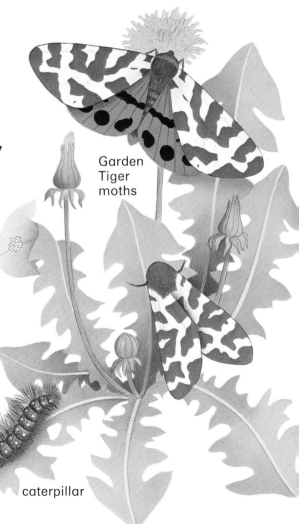

Garden
Tiger
moths

eggs

caterpillar

These moths belong to the same family
as the Garden Tiger and the Cinnabar.
The hairy caterpillars of the **White Ermine**
and **Wood Tiger** feed on many different
low growing plants.

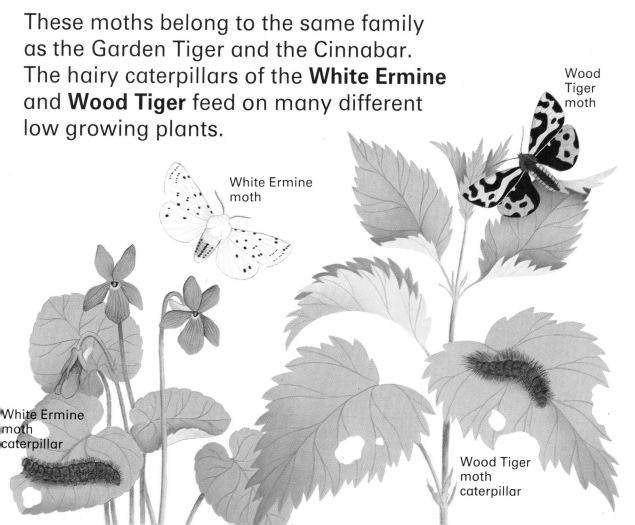

Wood
Tiger
moth

White Ermine
moth

White Ermine
moth
caterpillar

Wood Tiger
moth
caterpillar

There are seven different **Burnet moths**
but the Six Spot is the most common.
Its caterpillars hatch from yellow eggs and
feed on bird's-foot trefoil. They hibernate
for the winter and continued feeding in the spring.
They spin papery, boat-shaped cocoons attached
to grass stems. The moths come out
of the cocoon in July.
They fly in the daytime.

Six Spot
Burnet moth

cocoon

caterpillar

Although most caterpillars do no harm, some are pests and upset us if we find them. When you cut open an apple and find a maggot, it is actually the caterpillar of the **Codlin moth** which eats its way into the centre of the apple to feed on the pips and flesh. When it is ready to pupate, it leaves the apple, which has now probably fallen from the tree and finds a hiding place under the bark of the nearest tree. The moth will hatch the following summer.

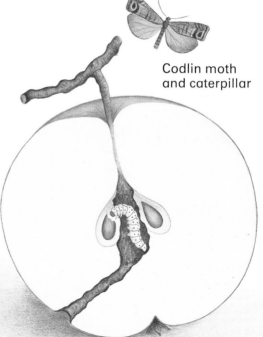

Codlin moth and caterpillar

There are several sorts of house moths and
clothes moths whose caterpillars eat wool and
feathers. They may remain in the caterpillar stage
for up to four years if they are short of food.
They used to do a lot of damage to clothes and carpets,
but now so many materials contain synthetics
which they can't eat, they are less of a pest.

NOTES ON KEEPING CATERPILLARS

Caterpillars can be kept in a jar or a transparent plastic box. Use a piece of fine netting as a lid so they cannot escape. Better still, is a wooden-framed cage with panels of fine netting. Caterpillars need fresh food each day, and they prefer young plants. Keep the plants fresh in a jar of water, plug the top of the jar with cotton wool so the caterpillars can't fall in and drown.

If your cage is big enough you can keep the food plant fresh for longer by growing it in a flower pot in the cage. As the caterpillars grow and eat more, they will need a new plant every few days. At the same time as you change the plant, clean out the cage, so that the caterpillars don't catch any disease. Be careful not to squash caterpillars when moving them. It is a good idea to pick up and move very tiny ones on the hair tip of a soft paint brush.

Caterpillars may pupate amongst the leaves of their food or against the side of their cage. Some species may pupate under soil, so you will need a layer of earth at the bottom of the cage.

If the caterpillars change into pupae in late Autumn they will need to be kept cool during the winter months so that they don't dry out. They can be kept in a sealed plastic box in a shed or even at the bottom of a refrigerator.